A MEDICAL REVOLUTION
FOR THE AGES

JULIAN LIEB, M.D

"When you have two competing theories which make exactly the same predictions, the one that is simpler is better."

Ockham's Razor

PREFACE

The ability to stimulate immune function holds the key to the future of human, veterinary, avian, marine and space medicine, as well as agriculture. Touted as unavailable, it was available in 1980, when Dr Louis Sherman showed that lithium stimulates immune function and prevents recurrences of bacterial skin infections, and in 1981 when I published the first of nine review articles on the immunostimulating and antimicrobial properties of lithium and antidepressants. Suppression of this innovation by vested interests has inflicted colossal human and economic losses on society, as Francis Bacon predicted it would.

Suppression of prostaglandins in 1990, in preference to biotechnology and genomics, inflicted another round of colossal human and economic losses. The pandemic commercialization of medicine is precisely what Bacon and Claude Bernard knew would invite disaster, for to innovate conflicts of interest must be avoided, to allow a scientist to think his own thoughts, be his own master, and follow his own conscience, Arthur Koestler noting that creative people are teacher and student in the same body. Bacon taught that to succeed in research one must practice more "humanity" than one's peers, humanity standing for ethics and charity, or altruism. Bacon and Bernard realized that inductive reasoning is a key to innovation, suppressed in preference to deductive since 1960.

Bacon notes, "There are and can be only two ways of searching into and discovering truth. The one flies from the senses and particulars to the most general axioms, and from these principles, the truth of which it takes for settled and immoveable, proceeds to judgment and to the discovery of middle axioms. And this way is now in fashion. *The other derives axioms from the senses and particulars, rising by a gradual and unbroken ascent, so that it arrives at the most general axioms last of all. This is the true way, but as yet untried.* Viscount Bacon, my innovations had their roots in the renaissance, and to the best of my knowledge are true.

TABLE OF CONTENTS

Please consult a biomedical database
for references.

CHAPTER ONE

QUOTATIONS FROM THE MEDICAL AND ETHICAL PHILOSOPHIES OF BYGONE ERAS

"Truth is the daughter of time."

Leonardo da Vinci

"Truth, not authority, is the daughter of time."

Francis Bacon

"If the misery of the poor be caused not by the laws of nature, but by our institutions, great is our sin."

Charles Darwin

"Whenever we depart from the great principles of truth and honesty, of equal freedom and justice to all men whether in our relations with other states, or in our dealings with our fellow-men, the evil that we do surely comes back to us, and the suffering and poverty and crime of which we are the direct or indirect causes, help to impoverish ourselves." (from Bad Times, 1885, as published by "The Alfred Russel Wallace Page)."

Alfred Russel Wallace

"AS THE births of living creatures, at first are ill-shapen, so are all innovations, which are the births of time. Yet notwithstanding, as those that first bring honor into their family, are commonly more worthy than most that succeed, so the first precedent (if it be good) is seldom attained by imitation. For ill, to man's nature, as it stands perverted, hath a natural motion, strongest in continuance; but good, as a forced motion, strongest at first. Surely every medicine is an innovation; and he that will not apply new remedies, must expect new evils; for time is the greatest innovator; and if time of course alter things to the worse, and wisdom and counsel shall not alter them to the better, what shall be the end? It is true, that what is settled by custom, though it be not good, yet at least it is fit; and those things which have long gone together, are, as it were, confederate within themselves; whereas new things piece not so well; but though they help by their utility, yet they trouble by their inconformity. Besides, they are like strangers; more admired, and less favored. All this is true, if time stood still; which contrariwise moveth so round, that a froward retention of custom, is as turbulent a thing as an innovation; and they that reverence too much old times, are but a scorn to the new. It were good, therefore, that men in their innovations would follow the example of time itself; which indeed innovateth greatly, but quietly, by degrees scarce to be perceived. For otherwise, whatsoever is new is unlooked for; and ever it mends some, and pairs others; and he that is holpen, takes it for a fortune, and thanks the time; and he that is hurt, for a wrong, and imputeth it to the author. It is good also, not to try experiments in states, except the necessity be urgent, or the utility evident; and well to beware, that it be the reformation, that draweth on the change, and not the desire of change, that pretendeth the reformation. And lastly, that the novelty, though it be not rejected, yet be held for a suspect; and, as the Scripture saith, that we make a stand upon the ancient way, and then look about us, and discover what is the straight and right way, and so to walk in it."

Francis Bacon, from "Advancements of Learning"

The Declaration of Independence reminds us that there are standards outside and above the teachings of governments, standards superior even to what a government might at any moment believe or choose. The right of revolution implies, that is, the supremacy of reason in human affairs.

The Duke of Wellington anticipated the damaging impact of peer review and editorial bias when remarking,

"I mistrust the judgment of every man in a case in which his own wishes are concerned."

"An Introduction to the Study of Experimental Medicine."

In "An Introduction to the Study of Experimental Medicine (1865), Claude Bernard describes what makes a scientific theory good, and what makes a scientist important, a true discoverer.

"When we meet a fact which contradicts a prevailing theory, we must accept the fact and abandon the theory, even when the theory is supported by great names and generally accepted."

"….. Proof that a given condition always precedes or accompanies a phenomenon does not warrant concluding with certainty that a given condition is the immediate cause of that phenomenon. It must still be established that when this condition is removed, the phenomenon will no longer appear…."

. "The dominant idea of despisers of their fellows is to find others' theories faulty and try to contradict them"….. (They) falsify science and the facts… lack an ardent desire for knowledge ….a man of science rises ever, in seeking truth; and if he never finds it in its wholeness, he discovers nevertheless very significant fragments; and these fragments of universal truth are precisely what constitutes science."

"In teaching man, experimental science results in lessening his pride more and more by proving to him every day that primary causes, like the objective reality of things, will be hidden from him forever and that he can only know relations."

"The idea is what establishes, as we shall see, the starting point or the primum movens of all scientific reasoning, and it is also the goal in the mind's aspiration toward the unknown."

Claude Bernard

"Paradoxes play a key role in the advancement of science. They are associated with excitement, and with the knowledge that we must be looking at something the wrong way. The clear formulation of a paradox can herald important advances, since the resolution of the paradox is generally a conceptual step forward. It is therefore of paramount importance to identify paradoxes and focus attention on them. A scientific paradox exists when there is a conflict between some well supported and widely accepted theoretical dogma or framework, and some piece of experimental or observational data."

Hoffman GW

Levy J G

Nepom GT

"Paradoxes In Immunology."

CHAPTER TWO

PROSTAGLANDINS: AN ESSENTIAL BUT SUPPRESSED MEDICAL INNOVATION

"AS THE births of living creatures, at first are ill-shapen, so are all innovations, which are the births of time. Yet notwithstanding, as those that first bring honor into their family, are commonly more worthy than most that succeed, so the first precedent (if it be good) is seldom attained by imitation. For ill, to man's nature, as it stands perverted, hath a natural motion, strongest in continuance; but good, as a forced motion, strongest at first. Surely every medicine is an innovation; and he that will not apply new remedies must expect new evils; for time is the greatest innovator; and if time of course alters things to the worse, and wisdom and counsel shall not alter them to the better, what shall be the end?"

Francis Bacon

The appalling quality of healthcare is the end, prostaglandins the corrupted innovation. Prostaglandins are ephemeral, infinitesimal, and powerful signaling molecules, regulating every physiological process and microanatomical structure of every cell; when up-regulated, physiology becomes pathology, the variations determined by genes. Prostaglandins are composed of carbon, oxygen, and hydrogen, with a configuration resembling a hairpin. Enter, "prostaglandins" in a biomedical database, and you will find thousands of studies revealing their regulation of our physiology, a physiological process not regulated by prostaglandins yet to be identified.

In experiments conducted in the department of obstetrics at the Columbia Medical School in the 1920's, Raphael Kurzrock and Charles Lieb noticed that when they attempted artificial insemination the uterus often expelled the semen. They found that human seminal fluid could affect the state of contraction of strips of muscle from the uterus, either contracting or relaxing them. They remarked in a paper published in 1930 that the history of the patients from whom the muscle strips were obtained made their experiments even more intriguing. Muscle from patients with a history of successful pregnancy responded to semen by relaxing, while semen always induced contractions in uterine muscle from women with a history of long acting sterility.

In the early 1930's, Maurice Goldblatt in the United States and Ulf von Euler in Sweden showed that factors in the seminal fluid of boars act on various smooth muscles and lower blood pressure. Von Euler named these substances "prostaglandins" because the prostate contains small amounts of them, and he assumed that what he had extracted from semen must have come from that gland. Today we know that every cell in our bodies, without exception, manufactures prostaglandins or other members of the eicosanoid family. Prostaglandins constitute a response system to stimuli. They orchestrate cognitive, emotional, behavioral physiological, pathological and reproductive responses to the environment.

At the Karolinska Institute in Stockholm, Sune Bergstrom purified several prostaglandins, determined their chemical structure, and showed that they are formed from essential fatty acids. After collaborating with Bergstrom from 1959-1962 on the structure of prostaglandins, Bengt Samuelsson provided a detailed picture of arachidonic acid and prostaglandin metabolism, and defined the chemical processes involved in their synthesis and breakdown. Samuelsson showed that platelets convert arachidonic acid to thromboxanes (TX's), while white blood cells convert it to leukotrienes (LT's). Thromboxanes constrict blood vessels and cause platelets to clump together and release more clotting factors. This is useful when clotting is necessary to stop bleeding; when this mechanism is overactive, it plays a pivotal role in heart attacks and strokes.

And thromboxanes directly stimulate the smooth muscles of blood vessels to contract, including those of the heart and brain.

At Oxford University in the mid-sixties, Sir John Vane and his colleagues developed the cascade super fusion bioassay technique for measurement of the release and fate of vasoactive hormones in the circulation or in the perfusion fluid of isolated organs. In 1971 Vane made the fundamental discovery that anti-inflammatory compounds such as aspirin block the formation of prostaglandins and thromboxanes. Later Vane and Salvador Moncada isolated a prostaglandin in the wall of blood vessels and named it prostacyclin (PGI2). In dilating blood vessels and inhibiting the aggregation of platelets, prostacyclin opposes the actions of thromboxanes. In some countries patients with such vasoconstrictive disorders as Raynaud's disease, peripheral vascular disease and pulmonary hypertension, whether spontaneous or caused by certain appetite suppressants, are treated with infusions of synthetic prostacyclin. When Dr. D Ansell infused a prostacyclin into four patients with vasoconstrictive disease, they all became depressed; when injected into mice, it induced a depression-like condition. For their pioneering research on prostaglandins, Bergstrom, Samuelson and Vane were awarded the Nobel Prize in Medicine in 1982.

The preimplantation human embryo would be destroyed by an attack by the maternal immune system, as half of its contents come from its father. Instead, the

embryo temporarily increases its production of prostaglandin E2, which, when elevated, becomes a powerful
immunosuppressant to ward off the attack. In the absence
of this mechanism, reproduction would be impossible. As
we shall see, prostaglandin E2 is involved in creating life
and destroying it, as prostaglandins are inherently paradoxical.

If the biological clocks in enzymes that produce prostaglandin E2 slowly increase its production, we will age
slowly, and unlikely to develop the numerous prostaglandin E2- generated disorders, that can shorten our lifespan,
including infectious and cardiovascular disorders, cancer,
neurodegenerative and autoimmune disorders, mood disorders, and others. Centenarians are virtually immune to
major medical and surgical disorders, their prostaglandins
on their best behavior.

Prostaglandins determine tolerance or intolerance
towards everything with which the body comes into
contact, among them temperature fluctuations, venom,
emotional stress, shear stress, microgravity, pathogens,
oxidative stress, changes in ionic composition, chrono-
tropic agonists, medications, alcohol, allergens, carcino-
gens including tobacco, and euphoriants, People gifted
with regulated prostaglandin production are relatively
immune to stress, people lacking such control hypersen-
sitive, with those with depression at the top of the list.
Virtually every medicine ever invented brushes with prostaglandins during its tour of duty. Aspirin and ibuprofen

are among those generally appreciated as inhibiting prostaglandins are aspirin and ibuprofen, lithium and antidepressants virtually unknown.

Databases contain copious evidence of the role of prostaglandins in virtually all of our diseases and disorders. .Antidepressants, which inhibit prostaglandins are variably effective for among others psoriasis, plantar warts, arthritis, hives, ciguatera fish poisoning, asthma, peptic ulcers, canker sores, cold sores, migraine, hiccup, fibromyalgia, multiple sclerosis, Parkinson's, Huntington's, and Alzheimer's diseases, shingles and cancer. This constellation can be explained only by down- regulation of prostaglandin E2 in the brain. This is a resisted paradigm shift that upon implementation promises to radically improve the quality of care thus slashing its cost.

In 1973, David Horrobin showed that lithium and antidepressants inhibit prostaglandins, and in 1977 that prostaglandins regulate the nucleic acids deoxyribonucleic (DNA) and ribonucleic acid (RNA). Subsequently, others showed that prostaglandins regulate the synthesis, inhibition, and expression of genes and the growth, differentiation, and replication of cells, with cancer the .accelerated replication of abnormal cells. Twenty years ago, a paradigm shift was in place that could have revolutionized the prevention and treatment of cancer.

In the nineteen-seventies-and-eighties, prostaglandins attracted substantial drug company investment, one of which referred to them as "Medicine's New Frontier." With new technology that accelerates the production of DNA, venture capitalists, the U.S patent office, medical schools and the media launched biotechnology, stampeding many drug companies into dropping prostaglandins to free up capital to purchase biotechnology companies and their patents. In 1980, the first genomics company was launched under the banner of a virtual cure all for all ailments, plausible to the uninformed but leaving the informed aghast at how genes could be accessed for clinical purposes. Proteinomics and stem cells followed, prostaglandins virtually disappearing from sight and mind even though they regulate proteins and stem cells..

Medical research is largely based on the premise that DNA and RNA in the cell nucleus, and enzymes and proteins in the cell, whose structures are defined by DNA, are of overwhelming importance in disease. While significant, their practical importance has been exaggerated. Membrane lipids, which modulate the behavior of these entities, offer far more opportunities for practical therapeutic interventions.

The idea that antidepressants might be effective for cancer was first explored fifty years ago, and ample proof has emerged. More than one hundred published studies have shown that antidepressants kill cancer cells, inhibit their proliferation, augment chemotherapy, protect non-malignant cells from ionizing radiation and chemotherapy toxicity, and convert multidrug resistant cells to sensitive.

Antidepressants can arrest cancer even in advanced stages, occasionally eradicate it, and significantly extend life. To verify one may access Medline or Pubmed, and enter "antidepressants" and "cancer."

Thirty years ago, Rashida Karmali showed that prostaglandins regulate the initiation, promotion, and spread of tumors. In 1973, David Horrobin was among those showing that lithium and antidepressants oppose prostaglandins, and in 1977, that prostaglandins regulate nucleic acids (DNA and RNA). Millie Hughes- Fulford and others followed by showing that prostaglandins regulate the synthesis, inhibition, and expression of genes. Later, Armato and Andreis showed that prostaglandins regulate the growth and differentiation of cells and Goodlad cell division, when cancer is the accelerated division of abnormal cells. Other prostaglandin inhibiting agents such as aspirin and ibuprofen have also shown considerable promise in preventing, treating, and arresting cancer..

Increased synthesis of prostaglandin E2, or inhibition of its degradation, beyond a critical threshold by brain enzymes, is responsible for depression on the one hand, and for defective immunity, autoimmunity, cardiovascular, and neurodegenerative disorders on the other. Antidepressants are often effective in preventing, alleviating, or reversing these disorders. They inhibit the enzymes that synthesize prostaglandins, and/or stimulate the enzyme that degrades them. These mechanisms illuminate what causes these disorders, but not their variations. Why does one person suffer from depression and heart disease, another from depression and recurrent infections, a third from depression and various autoimmune disorders? The answer surely lies with the nucleic acids that compose genes. Why would excessive production of prostaglandins in the brain account for defective immunity and autoimmunity? A provisional answer is that enzymes behave paradoxically. Why do antidepressants alleviate disorders of defective immunity and autoimmunity? Antidepressants have paradoxical actions on prostaglandins. Lithium inhibits the turnover of arachidonic acid by down regulating brain phospholipases, and thus has potent immunostimulating, antiviral, and antibacterial actions.

When the human genome was found to contain fewer genes than could account for our diversity, geneticists conjured up "epigenetics" to preserve the dominance of genes, while ignoring the earlier studies of Horrobin and Hughes-Fulford.

People not given to depression generally have stable prostaglandin E2 that increases slowly with time, many living to a ripe old age as they seldom get sick. It is those at risk for depression or have defective immune function from other causes, that are vulnerable to major illness, either spontaneously or under stress.. Increased prostaglandin E2 production causes depression, genes the variations.

The dark ages ended when Leonardo da Vinci showed his innovative prowess in public, inspiring other budding creative geniuses into joining him. A hundred years later, Francis Bacon realized that scientific innovation would release healthcare choices from tyranny, designed the scientific method to innovate, and emphasized the role of ethics and altruism in innovation and its implementation. Over the past forty years, pharma, politicians, the U.S patent office, Food and Drug Administration, National Institutes of Health, National Institute of Mental Health, Center for Disease Control, National Institute of Allergy and Infectious Disorders, medical school deans, tenured faculty, university presidents, Wall street, business executives, hypocritical ethicists, mediocre epidemiologists, medical and lay media, economists, statisticians, government, governors, all vested interests, siphoned control of one's health back to the "authority" Francis Bacon detested.and fought against by innovating the scientific method.

After failing for 31 years to persuade providers and recipients to adopt the innovations, I hope, "it be the reformation that draweth on the change, and not the desire of change, that pretendeth the reformation. And lastly, that the novelty, though it be not rejected, yet be held for a suspect; and, as the Scripture saith, that we make a stand upon the ancient way, and then look about us, and discover what is the straight and right way, and so to walk in it".

The innovative scientific paradox of treating the depression rather than what it predisposes to would be immensely cost effective, and liberate humanity from the shackles of the gargantuan tyrant referred to as health-care, as to restore it to the freedom envisaged by Francis Bacon, as well as society from the economic burden.

Thomas Huxley remarked earlier, "One small fact could ruin the grandest hypothesis." Genomics became what Bruce Charlton aptly refers to as "zombie science' science that died, but repeatedly resurrected by powerful car-tels because it stifles innovation. It would be a mistake to believe that equal or superior remedies are in the wings.32 years after the initial publication of immunostimulation, an investment of scores of billions of dollars in genomics achieved little, or to be embarrassed by the prospect of taking mood regulators, at the sacrifice of one's health and life.

In "Against Method" Paul Feyerabend noted that sup-pressing a paradigm in preference to one politically favored could permanently damage society. He cautioned that the guardians of paradigm failures seldom concede to valid newcomers, to the extent that political interven-tion might hold the only hope of progress. We are amazed at how easily people lost their minds in signing up for the South Seas bubble and the Dutch tulip mania; posterity will view us as losing ours by purging prostaglandins.

CHAPTER THREE

ONLY MINUTES TO MIDNIGHT

Depression predisposes to immune, cardiovascular, neurological, gastrointestinal, genitourinary, reproductive and endocrine disorders, the variations determined by genes.. The disorders include diabetes mellitus, Parkinson's, Alzheimer's and Huntington's diseases, seizure disorders, chronic obstructive lung disease, heart attacks, strokes, congestive heart failure and cancer. The intensity of depression appears to determine the severity, morbidity, and mortality of the disorders to which it predisposes. It is no longer a question of which disorders depression predisposes to, but to which it does not.

Increased synthesis of prostaglandins by brain enzymes above a critical threshold is responsible for depression on the one hand, and for defective immunity, autoimmunity, cancer, cardiovascular, and neurodegenerative disorders on the other. Antidepressants are often effective in preventing, alleviating or reversing these disorders. They inhibit the enzymes that synthesize prostaglandins, and stimulate the enzyme that degrades them. These mechanisms illuminate the cause of these disorders, but not their variations. Why does one person suffer from depression and heart disease, another from depression and recurrent infections, a third from depression and various autoimmune disorders? To what may one ascribe the variation among cancers? Cumulative evidence shows clearly that genes are responsible for variations. Antidepressants are like a clutch, disengaging the brain in overdrive from genes, thus restoring physiological activity.

Claude Bernard noted that applying mathematical standards to biology is ineffective because biology is too complex. The only certain method is removing the cause of a problem and observing the result. Depression as the cause of a disorder started as a trickle became a stream, and recently a flood, combined with the observation of disorders such as cancer, autism, cerebral palsy and Down's syndrome responding to antidepressants. The list of disorders not predisposed to depression, or nor responding to an antidepressant Is a small one, creating an ideal setting for applying Bernard's test of causality; we should treat depression, not its sequelae

Medical uses of antidepressants

"Make everything as simple as possible, but not simpler."

Albert Einstein

The following is an overview. Readers may search bio-medical databases for the status of disorders not listed here.

PREVENTION AND TREATMENT OF:

Cardiovascular disorders

High blood pressure (hypertension)

Heart attacks

Raynaud's disease

Arrhythmias

Congestive heart failure

Strokes

Peripheral arterial disease

Pulmonary emboli

Allergic disorders

Asthma

Allergic rhinitis (hay fever)

Urticaria (hives)

Ciguatera fish poisoning

Autoerythrocyte sensitization

Respiratory Disorders

Recurrent sinusitis

Recurrent sinobronchitis

Asthma

Tuberculosis

Sarcoidosis

Cancer

Gastrointestinal Disorders

Peptic ulcer disease

Irritable bowel syndrome

Chronic diarrhea

Recurrent canker sores

Recurrent cold sores

Ulcerative colitis

Crohn's disease

Cancer

Major Neurological Disorders

Cancer of the brain

Parkinson's disease

Alzheimer's disease

Lou Gehrig's disease (ALS)

Huntington's disease

Multiple sclerosis

Migraine

Tremor

Tinnitus

Stroke

Traumatic brain injury

Neurodegenerative Disorders

Alzheimer's disease

Multiple sclerosis

Parkinson's disease

Lou Gehrig's disease

Huntington's disease

MUSCULAR DYSTROPHY

An emerging field

DEVELOPMENTAL DISORDERS

Down's syndrome

Mental retardation

Cerebral palsy

Autism spectrum disorder

Pain

Low back pain

Arthritis

Shingles

Tic doloroux (trigeminal neuralgia).

Peripheral neuritis of diabetes

Cancer

Angina

Virtually any pain disorder

Genitourinary Disorders

Interstitial cystitis

Vulvodynia

Prostatitis

Bed-wetting

Soiling

Infertility

Recurrent spontaneous first trimester miscarriages

Recurrent yeast infections.

Premenstrual syndrome

Skin Disorders

Neurodermatitis

Psoriasis

Warts

Plantar warts

Cancer

Autoimmunity

In autoimmunity antibodies or immune cells react against some aspect of the self. Autoimmune diseases are common, and can cause substantial damage. DNA, viruses, bacteria and stress have been proposed as causing auto-immune disorders, but prostaglandins are undoubtedly key factors. Rheumatoid arthritis, migraine, asthma, multiple sclerosis, osteoporosis, pernicious anemia, and juvenile, insulin –dependent diabetes, are widely accepted as autoimmune disorders, and prostaglandins play a cardinal role in all.

Lithium and antidepressants can act on the brain so as to stimulate immune function or, paradoxically, a hyper immune (autoimmune) state. Thus prostaglandins are capable of acting paradoxically on the brain.

Infectious Disorders

Lithium

Numerous acute, bacterial, and viral disorders. I would favor lithium for an avian influenza pandemic, cholera; hospital- acquired infectious disorders, methicillin resistant staphylococcal infectious disorders (MRSA) and food poisoning.

Acute viral infectious disorders

Urinary tract disorders

Chalazions

Methicillin- resistant infectious disorders MRSA*

Hospital- acquired infectious disorders (HAIs).*

Salmonella*

E Coli*

Listeria*

Klebsiella pneumonia*

Legionnaire's disease*

Cholera*

* No data available as yet, but given lithium's immunos-timulating, antiviral and antibacterial properties, informed consent for patients would be mandatory.

Antidepressants

A plethora of bacterial, viral, fungal, and parasitic disorders including the acquired human immunodeficiency disor-der, tuberculosis and malaria.

CHAPTER FOUR

THE REMARKABLE ANTICANCER PROPERTIES OF ANTIDEPRESSANTS

In 1981, one year before the emergence of AIDS, I published the first of nine reviews on the immunostimulating and antimicrobial properties of lithium and antidepressants (1-8). Publication of many of these reviews was delayed by the intransigence of journal editors and reviewers, the editor of a premier medical journal rejecting one review for being, "too long." He and many of his peers acted as if effective treatments and preventives for AIDS were readily available, at a time when any idea was better than no idea. Fifteen years ago Jonas Salk conceded that a vaccine for AIDS was technically impossible, such that research should focus on trying to develop immunostimulants.

In August 2008 Anthony Fauci, the director of the Infection and Allergy division of the National Institutes of

Health, announced at the International AIDS conference in Mexico City, that attempts at developing an AIDS vaccine had failed, such that research should focus on trying to develop immunostimulants, without referring to Salk, nor to the articles I had mailed to NIH. It would seem that at NIH, literature searches are passé. Salk refused to patent the polio vaccine, but wished to see it become available to human beings everywhere.

His generosity of spirit surely played a role in his remarkable success as a virologist, in contradistinction to the self-interests of venture capitalists, drug companies, and the Harvard Business School. Immunostimulation is probably relevant to all pathogens, and to cancer prevention and treatment. Its potential value to mankind in both humanitarian and economic terms would have to run in the hundreds of billions of dollars. Inexpensive, generic agents defy commercial exploitation, a fortune to be saved, but not made.

In 2001 I published the first of four review articles on the potent anticancer properties of antidepressants (9-12). All were delayed by the suppressive tactics of editors and reviewers of many cancer and general medical journals. The National Cancer Institute and the American Cancer Society declined interest, as did scores of cancer foundations, societies, and organizations. Over the past thirty years I have written to thousands of people and institutions about infection and cancer, including more than eighty medical schools in the U.S, U.K, Europe, India, Canada and

South Africa. Not one, including the most prestigious, was accommodating.

In "Against Method" Paul Feyerabend noted that suppressing a theory in preference to one politically favored could permanently damage society (13). Accordingly, the anti-prostaglandin, immunostimulating and antimicrobial properties of lithium and antidepressants, and the anticancer properties of antidepressants, have been suppressed in preference to genetics, biotechnology and stem cells. Thomas Huxley noted that, "One small fact can ruin the grandest hypothesis."

In 1979 David Horrobin and his coworkers showed that prostaglandins regulate nucleic acids (14), and researchers have subsequently shown that prostaglandins regulate the synthesis, inhibition and expression of genes (15). Cancer is accelerated division of abnormal cells: In 1990 Goodlad and coworkers showed that prostaglandins regulate cell division (16). Many advances have little immediate impact, one of the reasons advanced that only a few people had the specialized knowledge to recognize their significance. Medicine has become so narrowly specialized that physicians tend to concentrate their reading on their own specialties. The neglect of prostaglandins is reflected in the August 2007 Medical College Admission Test (MCAT) that did not contain a single question on the omnipresent signalers.

A therapeutic advance is reinforced when the pharmacological mechanisms are known. In this instance, excessive synthesis of prostaglandins in the brain depresses mood and immune function, and activates the mechanisms of carcinogenesis, while antidepressants have unique prostaglandin-inhibiting properties (11, 12). Scores of thousands of publications attest to the role of prostaglandins in carcinogenesis. Conventional prostaglandin inhibiting drugs such as sulindac, piroxicam, ibuprofen and indomethacin have proven anticancer actions in vitro and in vivo (17). Whether to use a prostaglandin- inhibitor or an antidepressant in an individual case is a subject for future research, but I would tend to favor trying antidepressants first, not the least for their impressive immunostimulating and antimicrobial properties.

More than one hundred laboratories have shown that antidepressants kill cancer cells, inhibit their proliferation, reverse multidrug resistance, and protect nonmalignant cells from ionizing radiation and chemotherapy toxicity (11, 12). As low molecular weight mitocans, antidepressants target the mitochondria of cancer cells, while sparing health cells. Depression predisposes to cancer (18), and accelerates the death of patients with cancer, (19) while antidepressants extend the lives of lung, and probably other cancer patients, and enhance chemotherapy (20). Antidepressants can lose their effectiveness (21), in which case others can be substituted. The emergence of paradoxical reactions (22), such as intensification of depression, pain or any other cancer related symptom would militate for discontinuing an antidepressant and substituting another, perhaps chemically unrelated agent.

Despite the volume of evidence, there are bound to be editors and reviewers who will resort to the subterfuge of, "Interesting, but needs more evidence." Clinicians ought not to wait for some third party to give its stamp of approval for using antidepressants in the prevention, treatment and palliation of cancer, or to at least inform their patients of the neglected resource. Many antidepressants may have to be tried before achieving success. In some instances antidepressants may be combined with chemotherapy or radiation, in others they alone will suffice. Antidepressants have the remarkable ability to arrest cancer, and then reverse it. Clinicians should consider submitting case reports, both positive and negative, to build a body of evidence about malignancies that respond or seem to be impervious to antidepressants.

In combating cancer and infection, clinicians will need a wider selection of antidepressants than currently available. Agencies such as the Food and Drug Administration must begin to view antidepressants as immunostimulants and anti-carcinogens, and develop more permissive policies to approve agents with antidepressant properties. The development of biological markers to match antidepressants and subjects would apply to every context in which antidepressants are used. In all likelihood, such markers will involve prostaglandins.

Vested interests have colluded in the suppression of the remarkable immunostimulating and anticancer properties of antidepressants (23, 24, 25, and 26). In the world of commerce this might be regarded as monopolistic, and subject to antitrust laws. Applying those laws to medical research, education and publishing would seem to be overdue.

REFERENCES:

1. Lieb J. Immunopotentiation And Inhibition Of Herpes Virus Activation During Therapy With Lithium Carbonate. Med Hypoth 1981 7:885-890

2. Lieb J. Remission of herpes virus infection and immunopotentiation with lithium carbonate: Inhibition of prostaglandin E1 synthesis by lithium may explain its antiviral, immunopotentiating, and antimanic properties. Proceedings of the Third World Congress of Biological Psychiatry, Stockholm. Biol Psych Perris C, Struwe G, Jansson B. (eds.) 1981 695-698

3. Lieb J. Remission of rheumatoid arthritis and other disorders of immunity in patients taking monoamine oxidase inhibitors. Int J Immunopharm 1983 5(4): 353-357

4. Lieb J. Lithium and immune function. Med Hypoth 1987 23:73-93

5. Lieb J. Invisible antivirals. Int J Immunopharm 1994 16:

6. Lieb, J. Lithium and antidepressants: inhibiting eicosanoids, stimulating immunity, and defeating microorganisms. Medical Hypotheses (2002) 59(4), 429-432

7. Lieb, J. The immunostimulating and antimicrobial properties of lithium and antidepressants. Journal of Infection (2004) 49(2) 88-93

8. Lieb, J Lithium and antidepressants: Stimulating immune function and preventing and reversing infection. Medical Hypotheses (2007) 69, 8-11

9. Lieb, J. Antidepressants, eicosanoids and the prevention and treatment of cancer. Plefa (2001) 65(5&6), 233-239

10. Lieb, J. Antidepressants, prostaglandins and the prevention and treatment of cancer. Medical Hypotheses (2007) 684-689

11. Lieb, J.The multifaceted value of antidepressants in cancer therapeutics. Editorial Comment. European Journal of Cancer 44 (2) 2008 172-174

12. Lieb, J.Defeating cancer with antidepressants. ecancermedicalscience DOI 10.3332/ eCMS.2008.88

13. Feyerabend, P "Against Method." 1988. London, Verso.

14. Horrobin, DF, Manku MS. Roles of prostaglandins suggested by the prostaglandin agonist/antagonist actions of local anaesthetic, anti-arrhythmic, anti-malarial, tricyclic anti-depressant and methyl xanthine compounds. Effects on membranes and on nucleic acid function. Med Hypotheses. 1997 Mar-Apr 3(2):71-86

15. Hughes-Fulford, M. Prostaglandin regulation of gene expression and growth in normal and malignant tissues. Adv Exp Med Biol 1997 400 A():269-78.

16. Goodlad RA, Madgwick AJ, Moffatt MR, Levin S, Allen JL, Wright, NA. The effects of the prostaglandin analogue, misoprostol, on cell proliferation and cell migration in the canine stomach. Digestion. 1990 46 Suppl 2:182-7

17. Sporn MB, Hong WK. Concomitant DFMO and sulindac chemoprevention of colorectal adenomas: a major clinical advance. Nature Clin Prac Oncol doi:10.1038/nepone 1221

18. Penninx B, Guralnik JM, Pahor M, et al. Chronically depressed mood and cancer risk in older patients. J Natl Cancer Inst 1998 Dec 16 90(24):1888-93.

19. Lloyd-Williams M, Shiels C, Taylor F, Dennis M. Depression: an independent predictor of early death in patients with advanced cancer. J Affect Disord. 2008. Jun 14, Epub ahead of print

20. Jovanovic D, Nagorni-Obradovic LJ, Blanka, A, Popevic S, Grujic M. Mianserin therapy in advanced lung cancer patients. 8th Central European Lung Cancer Conference, Vienna, 2002. Internat. Proc. Division, Monduzzi Editore 2002 339-343.

21. Lieb, J and Balter A . Antidepressant tachyphylaxis. Med Hypoth 1984 15: 279

22. Lieb, J. Variation: Darwin's finches, sea barnacles, and the side effects of antidepressants. (Editorial). Med Hypoth (2008) 70, 221-223

23. Horrobin DF.Something Rotten at the Core of Science? Trends in Pharmacological Sciences. (2001) Vol. 22, No 2, 1-4

24. Horrobin, DF. Effective Clinical Innovation: an ethical imperative. Lancet (2002) 359: 1857-58

25. Horrobin DF. Are large clinical trials in rapidly lethal diseases usually unethical? Lancet (2003) 361: 695-97

26. Horrobin, DF. Not in the genes: enthusiasts for genomics have corrupted scientific endeavor and undermined hopes of medical progress. The Guardian (Feb 12, 2003) 1-3

CHAPTER FIVE

STIMULATING IMMUNE FUNCTION TO DEFEAT MICROORGANISMS AND TOXINS

Stimulating defective immune function to perform efficiently is a desirable approach to defeating pathogens. Such stimulation is represented as unavailable, while in truth the immunostimulating properties of lithium and antidepressants were documented many years ago.[1-4] A therapeutic claim is reinforced when the mechanism is known. Prostaglandins, when produced excessively, depress every component of immune function, and induce microbial replication. Wherever HIV comes into contact with arachidonic acid, an envelope glycoprotein powerfully converts this precursor to prostaglandin E2, depressing immune function and promoting viral replication, excessive prostaglandin E2 a leading candidate for the immunosuppression that is the hallmark of AIDS.[5-7]Antidepressants inhibit the synthesis of prostaglandin E2, antagonize its actions, and stimulate the primary prostaglandin-degrading enzyme.[8-10]

Collective evidence shows that lithium has immuno-stimulating, antiviral, and antibacterial properties,[11] anti-depressants immunostimulating, antiviral, antibacterial,[1-4] anti-parasite, and fungicidal properties.[12-15] Lithium is often effective for bacterial skin infections, aphthous ulcers, cold sores, and genital herpes,[11] antidepressants for aphthous ulcers, cold sores, and genital herpes.[11] Tuberculosis, now the #1 killer of the HIV infected, is developing resistance to standard treatment. In the late nineteen forties, physicians working in tuberculosis sanitaria observed patients with elevations of mood and energy. Their charts revealed that all were taking the monoamine oxidase inhibitors isoniazid or iproniazid, an observation from which antidepressant therapy developed. If antituberculotics double as antide-pressants, surely antidepressants must double as antitu-berculotics? The antimalarial properties of antidepressants in vitro are supported by many studies.[12] When added to antiretrovirals, antidepressants can reduce HIV viral loads to undetectable.[16] The authors of this study attribute this to adherence, seemingly unaware of the antiviral proper-ties of antidepressants. The advantage of immunostimu-lation is its non-specificity, a stimulated immune system indifferent to antigenicity.

People with intact immune function are relatively invulnerable to pathogens and toxins, compared to those with defective function. Depression is a seldom mentioned cause of defective immunity, although indices of immune function indicate that it does so.[17] In a study of 405 HIV-positive gay and bisexual men, those who reported being depressed throughout the eight-year study period, were two-thirds more likely to die than those who were never significantly depressed.[18]

Forty years ago, prostaglandins were shown to regulate immune function, and lithium and antidepressants to inhibit prostaglandins. Gradually, prostaglandins were found to regulate every aspect of HIV replication, and HIV to stimulate prostaglandin E2 production, to a greater degree than other viruses. This prostaglandin, when produced excessively, is thought to be responsible for the immune depression that is the hallmark of AIDS. Twenty five years ago, I believed that lithium and antidepressants could be used as heavy artillery against HIV, but when lithium failed to improve patients with AIDS in two small clinical trials, came to favor antidepressants for this purpose.[19,20,21,22]

1. Lieb J. Remission of herpes virus infection and immunopotentiation with lithium carbonate: inhibition of prostaglandin E1 synthesis by lithium may explain its antiviral, immunopotentiating, and antimanic properties. Biol Psychiatry 1981; 695-698.

2. Lieb J. Remission of rheumatoid arthritis and other disorders of immunity in patients taking monoamine

oxidase inhibitors. Int J Immunopharmacol 1983; 5(4): 353-357.

3. Rosenthal S, Fitch W. The antiherpetic effects of phenelzine. J Clin Psychopharmacol 1987; 7(2):119.

4. Murphy D, Donnelly C, Moskowitz J. Inhibition by lithium of prostaglandin E1 and norepinephrine effects on cyclic adenosine monophosphate production in human platelets. Clin Pharmacol Ther 1973; 14(5):810-814.

5. Lee R. The influence of psychotropic drugs on prostaglandin biosynthesis. Prostaglandins 1974; 5(1):63-68.

6. Manku MS, Horrobin DF. Chloroquine, quinine, procaine, quinidine and clomipramine are prostaglandin agonists and antagonists. Prostaglandins 1976; 12: 789-801.

7. Mak O, Chen S. Effects of two antidepressant drugs imipramine and amitriptyline on the enzyme activity of 15-hydroxyprostaglandin dehydrogenase purified from brain, lung, liver and kidney of mouse. Prog Lipid Res 1986; 25: 153-155.

8. Fernandez-Cruz E, Gelpi E, Longo N, Gonzalez B, de la Morena, MT, Montes, MG, Rosello , J, Ramis I,Suarez A, Fernandez, A. Increased synthesis and production of prostaglandin E2 by monocytes from drug addicts with AIDS. AIDS 1989; 3: 93-96.

9. Wahl L, Corcoran M, Pyle S, Pyle SW, Arthur LO, Harel-Bellan A, Farrar WL. Human immunodeficiency virus

glycoprotein (gp120) induction of monocyte arachidonic acid metabolites and interleukin 1. Proc. Natl Acad. Sci. USA 1989; 86:621-625.

10. Dumais N, Barbeau B, Olivier M, Tremblay MJ. Prostaglandin E2 up-regulates HIV-1 long terminal repeat-driven gene activity in T cells via NF-kappa B-dependent and - independent signaling pathways. J Biol Chem 1998; 273(42): 27306-27314

11. Dutta P, Pinto J, Rivlin R. Antimalarial properties of imipramine and amitriptyline. J Protozool 1990; 37(1): 54-58.

12. Lieb,J."The immunostimulating and antimicrobial properties of lithium and antidepressants." J Infection (2004) 49; 88-93

13. Lass-Florl C, Dierich MP, Fuchs D, Semenitz E, Ledo-chowski M. Antifungal activity against Candida sp. by the selective serotonin reuptake inhibitor sertraline. Clin Infect Dis 2001; 33(12):E135-136.

14. Munoz-Bellido J, Munoz-Criado S, Garcia-Rodri-guez J. Antimicrobial activity of psychotropic drugs: selective serotonin reuptake inhibitors. Int J Antimicrob Agents 2000; 14(3): 177-180.

15. Tsai A, Weiser S, Petersen M, Ragland K, Bangsberg D. Effect of antidepressant medication treatment on ARV adherence and HIV-1 RNA viral load in HIV+ homeless and

marginally housed individuals. In: Program and abstracts of the 16th Conference on Retroviruses and Opportunistic Infections; February 8-11, 2009; Montreal, Canada. Abstract 584

16. Frank M, Hendricks S, Johnson D, Wiesler J L, Burke WJ. Antidepressants augment natural killer cell activity: in vivo and in vitro. Neuropsychobiology 1999; 39(1):18-24.

17. Mayne TJ, Vittinghoff E, Chesney MA, Barrett DC, Coates TJ. Depressive affect and survival among gay and bisexual men infected with HIV. Arch Intern Med. 1996 Oct 28; 156(19):2233-8.

18. Lieb,J."Stimulating immune function to kill viruses." (And bacteria, parasites, and fungi). 2009, Amazon

19. Evans DL, Ten Have TR, Douglas SD, Gettes DR, Morrison M, Chiappini MS, Brinker-Spence P, Job C, Mercer DE, Wang YL, Cruess D, Dube B, Dalen EA, Brown T, Bauer R, Petitto JMAssociation of depression with viral load, CD8 T lymphocytes, and natural killer cells in women with HIV infection. Am J Psychiatry. 2002 Oct; 159(10):1752-9.

20. Evans DL, Lynch KG, Benton T, Dube B, Gettes Tustin NB, Lai JP, Metsger D, Douglas SD Selective serotonin reuptake inhibitor and substance P antagonist enhancement of natural killer cell innate immunity in human immunodeficiency virus/ acquired immunodeficiency syndrome. Biol Psychiatry 2008 May 1:63(9):899-905. Epub 2007 Oct 22.

21. Benton T, Lynch K, Dube,B, Gettes DR, Tustin NB, Lai JP, Metsger DS, Blume J, Douglas SD, Evans DL. Selective Serotonin Reuptake Inhibitor Suppression of HIV Infectivity and Replication Psychosom Med 2010 Oct 14

22. Ren X, Meng F, Yin J, Li G, Li X, Wang C, Herrler G. Action mechanisms of lithium chloride on cell infection by transmissible gastroenteritis coronavirus. PLoS One. 2011 May 6; 6(5):e18669.

CHAPTER SIX

TREATMENT MADE SIMPLE

Upon stimulation, phospholipases (enzymes) release arachidonic acid from membrane phospholipids. Arachidonic acid enters its cascade, where cyclooxygenase converts it to prostaglandins, thromboxane synthase (in platelets) to thromboxanes, and lipoxygenases to leukotrienes. People physiologically equipped to withstand biotic and abiotic stress, are able to prevent an induction of these enzymes. People with defective enzyme function will experience an induction, the consequences determined by nucleic acids (genes).In these individuals, preventive or therapeutic use of lithium and antidepressants would preserve health, and offer freedom from healthcare systems.

Immunostimulation is nonspecific, thus of paramount importance for all infectious disorders. It is not amenable to clinical trials, which in any event would be unethical.

I would favor lithium for swine and bird flu infections, and for prevention in those that wish to take it for that purpose. Clinical observation will reveal whether lithium or antidepressants are to be preferred.

HUMAN RIGHTS

Whereas recognition of the inherent dignity and of the equal and inalienable rights of all members of the human family is the foundation of freedom, justice and peace in the world,

Whereas disregard and contempt for human rights have resulted in barbarous acts which have outraged the conscience of mankind, and the advent of a world in which human beings shall enjoy freedom of speech and belief and freedom from fear and want has been proclaimed as the highest aspiration of the common people,

Everyone has the right to freedom of opinion and expression; this right includes freedom to hold opinions without interference and to seek, receive and impart information and ideas through any media and regardless of frontiers.

The right to the highest attainable standard of health is codified in numerous legally binding international and regional human rights treaties. The right, or other health-related rights, is also enshrined in over 100 constitutions worldwide.

It is the ethical and human right of every patient, physician, and citizen to receive this information, allowing them to have the final say as to whether or not to take advantage of it. The right does not allow anyone to evaluate or delay its dissemination, through every channel of communication available locally, regionally, nationally and internationally.

This book is for educational purposes only. All treatment decisions to be made with a physician

www.ingramcontent.com/pod-product-compliance
Lightning Source LLC
Chambersburg PA
CBHW060227290526
45789CB00003B/1451